WILD BIOMES

ANIMALS OF THE AFRICAN SAVANNA

by Mari Schuh

PEBBLE
a capstone imprint

Pebble Emerge is published by Pebble, an imprint of Capstone.
1710 Roe Crest Drive
North Mankato, Minnesota 56003
www.capstonepub.com

Copyright © 2022 by Capstone. All rights reserved. No part of this publication may be reproduced in whole or in part, or stored in a retrieval system, or transmitted in any form or by any means, electronic, mechanical, photocopying, recording, or otherwise, without written permission of the publisher.

Library of Congress Cataloging-in-Publication Data
Names: Schuh, Mari C., 1975- author.
Title: Animals of the African savanna / by Mari Schuh.
Other titles: Wild biomes.
Description: North Mankato, Minnesota : Pebble Emerge, published by Pebble, an imprint of Capstone, 2022. | Series: Wild biomes | Includes bibliographical references and index. | Audience: Ages 6-8 | Audience: Grades 2-3 | Summary: "Biomes are home to unique animals and plants. Introduce beginning readers to the African Savanna biome! Readers will get an up-close look at the characteristics of the land and weather and how zebras, lions, giraffes, and other animals have adapted to life in this amazing biome"-- Provided by publisher.
Identifiers: LCCN 2020027287 (print) | LCCN 2020027288 (ebook) | ISBN 9781977131911 (library binding) | ISBN 9781977132932 (paperback) | ISBN 9781977153807 (pdf)
Subjects: LCSH: Savanna animals--Africa--Juvenile literature. | Savanna ecology--Africa--Juvenile literature.
Classification: LCC QL115.3 .S38 2022 (print) | LCC QL115.3 (ebook) | DDC 591.74/8096--dc23
LC record available at https://lccn.loc.gov/2020027287
LC ebook record available at https://lccn.loc.gov/2020027288

Image Credits
Capstone: Eric Gohl, 5; Shutterstock: Barbara Ash, 20, Bobby Bradley, 6, CECIL BO DZWOWA, 15, CherylRamalho, 13, Keith 316, 10, KUV Photography, 12, Maggy Meyer, Cover, MattiaATH, 7, NelisNienber, 17, nwdph, 14, Oleg Znamenskiy, 11, RMFerreira, 9, TheMumins, 1, TTphoto, 4, 19, Vladimir Sazonov, 21

Design Elements
Capstone; Shutterstock: TheMumins

Editorial Credits
Editor: Jaclyn Jaycox; Designer: Hilary Wacholz; Media Researcher: Jo Miller; Production Specialist: Spencer Rosio

All internet sites appearing in back matter were available and accurate when this book was sent to press.

Table of Contents

The African Savanna 4
Savanna Animals 6
Life on the Savanna 16

Finding Patterns 20
Glossary 22
Read More 23
Internet Sites 23
Index 24

Words in **bold** are in the glossary.

THE AFRICAN SAVANNA

Zebras nibble on grass in the African **savanna**. The savanna covers about half of Africa. The land is flat. Grass covers most of it. Few trees grow there. The savanna is warm all year. It has a rainy season and a dry season. Many animals live there.

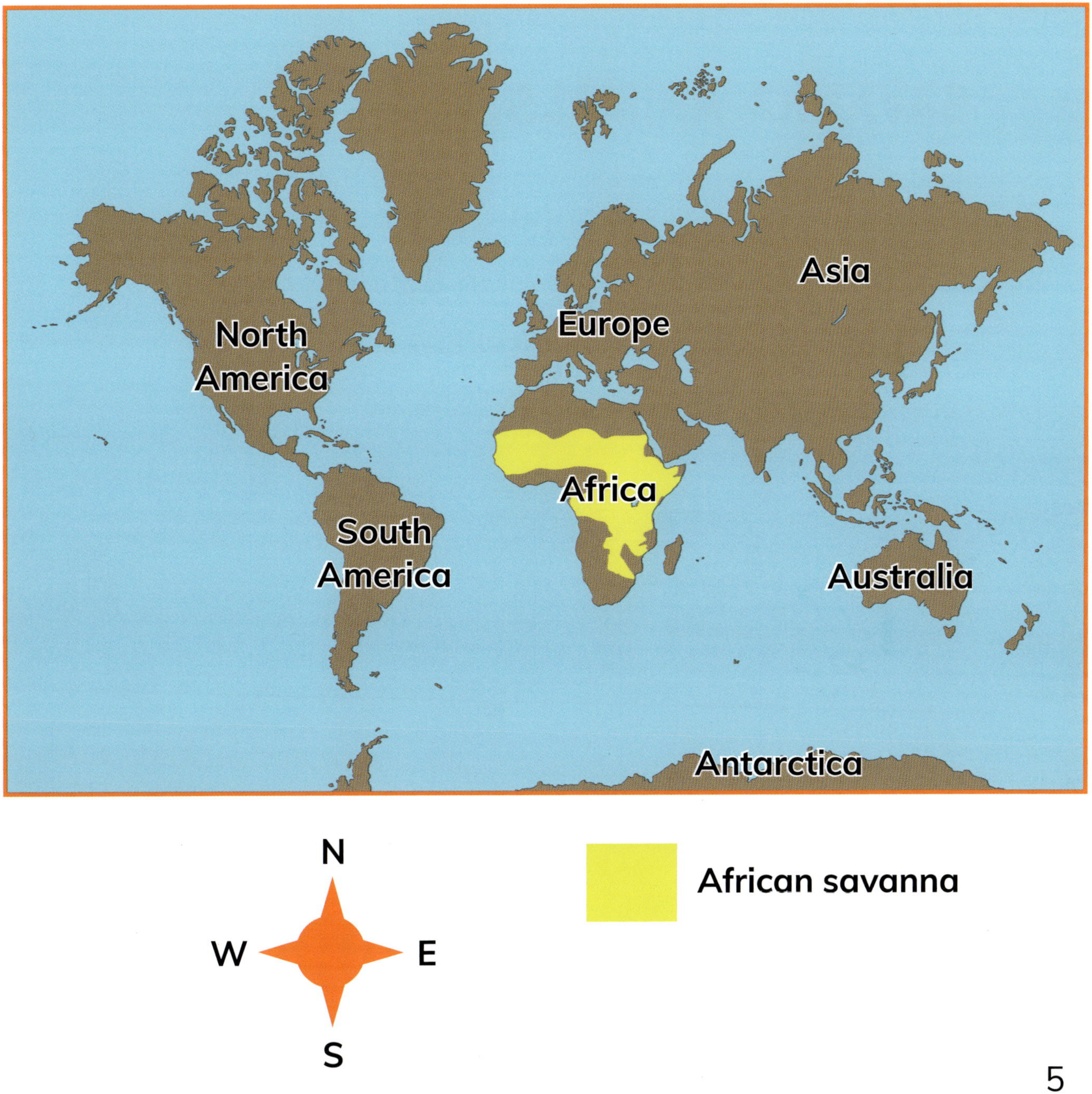
North America
Europe
Asia
Africa
South America
Australia
Antarctica
N
W
E
S
African savanna

SAVANNA ANIMALS

Lions live in the savanna. They hunt at night. They see very well in the dark. Female lions often hunt together. They eat wildebeests, zebras, and antelopes.

Look at the stripes! The stripes on zebras help them blend together. This keeps them safe from lions and other **predators**. It's hard for predators to pick out one to hunt.

Giraffes are the tallest animals on land. They can eat leaves on high tree branches. Giraffes can also see predators far away.

Giraffes have spotted coats. Their spots can make them hard to see. They blend in with tree leaves and shadows. This helps keep them safe.

Cheetahs usually hunt during the day. These fast hunters have spotted coats. The spots help them hide in tall grass. Cheetahs must eat **prey** quickly. Stronger animals might come to take it away.

Chomp! Wildebeests eat grass and leaves. They **migrate** when the rainy season ends. Thousands of wildebeests travel many miles. Zebras and gazelles migrate too. They look for food and water.

Elephants are big and strong. They push down trees. They eat the leaves, twigs, and bark. Sometimes water is hard to find. Elephants tear open tree trunks. They drink the tree's water.

Elephant **herds** keep young elephants safe. The young stay in the middle of the group. Lions and other predators can't reach them.

A gazelle runs across the savanna. Its long legs help it get away from predators. Lions, cheetahs, hyenas, and other animals hunt gazelles. Gazelles run away quickly!

Speed helps gazelles in other ways. Fires can burn in the savanna. Gazelles outrun the fires. They go to safe areas.

LIFE ON THE SAVANNA

Plants and animals in the savanna are part of a **food web**. Each animal in the food web is important. They need one another to survive. Lions need other animals to eat. Warthogs need plants to eat. Bugs are food for birds. Hyenas eat dead animals.

warthogs

Humans are a danger to savanna animals. Farmers raise **cattle**. Then there is less grass for wild animals. Farmers also clear land to grow **crops**. Wild animals have fewer areas to live. Many people are trying to protect the animals that live there.

Finding Patterns

Patterns on some animals help them stay safe and hide. Look at the photos in this book. Look at the fur and hair on the animals. What patterns do you see? Use crayons to draw and color two examples.

Glossary

cattle (KAT-uhl)—cows raised for dairy products or beef

crop (KROP)—a plant farmers grow in large amounts, usually for food

food web (FOOD WEB)—many food chains connected to one another

herd (HURD)—a large group of animals that lives or moves together

migrate (MYE-grate)—to move from one area to another

predator (PRED-uh-tur)—an animal that hunts other animals for food

prey (PRAY)—an animal that is hunted by another animal for food

savanna (suh-VAN-uh)—a flat, grassy area of land with few or no trees

Read More

Cosneau, Olivia. *Animals of the Savanna*. New York: Princeton Architectural Press, 2019.

Kortuem, Amy. *A Herd of Elephants*. North Mankato, MN: Pebble, 2020.

Pettiford, Rebecca. *Savanna Food Chains*. Minneapolis: Jump!, Inc., 2017.

Internet Sites

DK FindOut: African Savanna
www.dkfindout.com/us/animals-and-nature/habitats-and-ecosystems/african-savanna/

Easy Science for Kids: The Savanna
easyscienceforkids.com/the-savanna/

San Diego Zoo: Spectacular Savannas
kids.sandiegozoo.org/stories/spectacular-savannas

Index

antelopes, 6

bugs, 16

cattle, 18

elephants, 12–13

farmers, 18

giraffes, 8

food webs, 16

gazelles, 11, 14–15

grass, 4, 10, 11, 18

herds, 13

hyenas, 14, 16

leaves, 8, 11, 12

lions, 6, 7, 13, 14, 16

migrating, 11

predators, 7, 8, 13, 14

prey, 10

seasons, 4, 11

spots, 8, 10

stripes, 7

trees, 4, 8, 12

water, 11, 12

wildebeests, 6, 11

zebras, 4, 6, 7, 11